OVER ALL FRONTS

He 111

Franz Kober

Heinkel He 111 P-1 bombers of the III. *Gruppe* of
Kampfgeschwader 255.

SCHIFFER MILITARY HISTORY
West Chester, PA

Sources
VFW Fokker
Nowarra Archives
Podzun Archives
Heinkel, *Sturmisches Leben* (A Turbulent Life)
Nowarra-Kens, *Die deutschen Flugzeuge 1933-45* (German Aircraft 1933-45)

Foreword

The He 111 flew over every front from the first day of the Second World War to the last and was among the most produced aircraft of the Luftwaffe. Aerodynamically a very clean aircraft, it was a superior bomber during the first two years of the war. We have therefore decided to bring out a volume on this aircraft.

Translated from the German by David Johnston.

Copyright © 1991 by Schiffer Publishing Ltd.
Library of Congress Catalog Number: 91-60861.

Printed in the United States of America.
ISBN: 0-88740-313-1

This title was originally published under the title, *He 111 an allen Fronten,* by Podzun-Pallas-Verlag GmbH, 6360 Friedberg 3 (Dorheim). ISBN: 3-7909-0064-8.

We are interested in hearing from authors with book ideas on related topics.

Published by Schiffer Publishing, Ltd.
1469 Morstein Road
West Chester, Pennsylvania 19380
Please write for a free catalog.
This book may be purchased from the publisher.
Please include $2.00 postage.
Try your bookstore first.

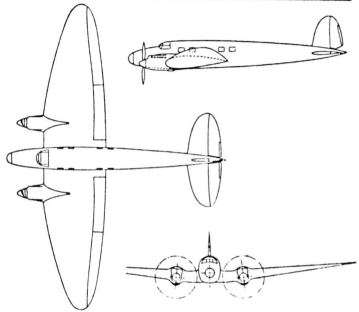

HEINKEL He 111
mit 2 BMW VI

Baujahr: 1933/34
Verwendungszweck: Post- und Passagierflugzeug für Schnellverkehr
Anzahl der Sitzplätze: 2 Mann Besatzung, 10 Passagiere
Besondere Einrichtungen: Einziehbares Fahrwerk, einziehbarer Sporn, Lande-klappen, Kabinenheizung, Kabinenbelüftung, Raucher- und Nichtraucher-Kabine, Toilette

Spannweite .... 22,6 m
Höhe .... 3,9 m
Länge .... 17,1 m

	BMW VI	Neue deutsche Hoch-Leistungsmotoren
Leergewicht	5300 kg	5420 kg
Fluggewicht	7870 kg	8000 kg
Reichweite	800—1500 km	1000—1500 km
Höchstgeschwindigkeit	345 km/h	410 km/h
Reisegeschwindigkeit	300 km/h	350 km/h
Landegeschwindigkeit	92 km/h	92 km/h

Anzahl, Lage und Fassungsvermögen der Kraftstoff- und Oeltanks:
2 Kraftstofftanks in den Tragflächen je 600 Ltr. .... zusammen **1200 Ltr.**
2 Oeltanks in den Tragflächen je 35 Ltr. bzw. 40 Ltr. .... 70 Ltr. bzw. 80 Ltr.

A three-view drawing of the He 111 C from 1938.

The Development of the He 111

The He 111 — a bomber, or *Kampfflugzeug* as it was known in official Luftwaffe parlance — was originally developed on the basis of a concept by the German airline *Lufthansa*, and was therefore conceived as a civil transport aircraft. It was due to the business sense of Ernst Heinkel that the aircraft became a bomber. Earning the trust of all who flew it, the He 111 was among the most reliable combat aircraft built in Germany.

Following the success of the Heinkel He 70 "Blitz," *Lufthansa* framed a requirement a similar type of aircraft offering greater seating capacity. In addition to the pilot and radio operator the aircraft was to carry ten passengers. Heinkel reckoned, however, that the few aircraft that *Lufthansa* would order would never cover the project's development costs. He therefore decided to have the design team of Walter and Siegfried Günter develop the aircraft from the outset so that it could become a bomber for the Luftwaffe. At that time the Luftwaffe was flying the Do 23 and Ju 52/3m bombers which were capable of 270 kph at best. Heinkel's high-speed He 70 being flown by *Lufthansa* was capable of 335 kph, and the current Luftwaffe fighters (He 51, Ar 65 and Ar 68), all biplanes, could reach 330 kph! The company's He 70 had already been built as a military reconnaissance aircraft because of its high-speed capabilities.

Although possessing the same lines as the He 70 the new aircraft was powered by twin engines. The engines were the same type used in the earlier aircraft: namely, the sole high-performance in-line engine Germany had at that time, the BMW VI of approximately 650 h.p. When the first aircraft of the new type, the He 111 A (later designated the He 111 V-1) was rolled out of the

An ancestor of the He 111 was the Bäumer B II/IV (above), the first design by the brothers Walter and Siegfried Günter, which attained a speed of 230 kph with an engine of only 65 h.p. The next development stage was the He 64 (below) which, flown by Hans Seidemann, was the sensation of the *Europa Rundflug* of 1932. Powered by an Argus As 8 R of 150 h.p., it had a maximum speed of 245 kph.

He 70/170

The He 70 (above) was developed to rival the American Lockheed "Orion" and, with a top speed of 377 kph, was some 50 kph faster. Developed from the *Lufthansa* He 70, the He 70 F reconnaissance aircraft (left) saw service in Spain from 1937.

Hungary received the He 170 (below), an He 70 with a French Gnome-Rhone K 14 engine.

The He 111 C first entered service as a civil transport with the airline *Lufthansa*. During the war most of these aircraft served as courier aircraft with the Luftwaffe.

The He 111 V 16 D-ASAR (above) was a special version. Powered by two DB 600 engines, it served as the personal transport of German State Secretary Erhard Milch.

He 111 V 5, D-APYS (below), was the prototype for the first bomber series, the B-1. The machine was later rebuilt and received the registration D-AJAK.

hangar in Rostock-Marienehe for the first time, everyone saw that, although the aircraft had a passenger cabin with windows, it also had a plexiglass nose with a machine-gun mount. In contrast to the mixed construction He 70, the He 111 was an all-metal aircraft, the first to be designed by Heinkel. On 24 February 1935 Heinkel's chief pilot, Gerhard Nitschke, took the He 111 into the air for the first time. Afterwards he declared himself very satisfied with the flight. In particular, he had praise for the aircraft's outstanding landing characteristics, which were later to endear the He 111 to the pilots of the Luftwaffe.

In developing the He 111 as a bomber and civil transport aircraft, Heinkel was in competition with the Junkers firm and its Ju 86, which was of a similar concept. Although the Ju 86 did enter production, it never equalled the success of the He 111. Heinkel now built a prototype for *Lufthansa*, the He 111 V-2 (D-ALIX), and one for the Luftwaffe, the He 111 V-3 (D-ALES). The definitive configuration for the *Lufthansa* aircraft was achieved in the V-4 (D-AHAO), and five similar machines were built as the C-O series. This was followed by the G-O, which featured a new wing, and the G-3, which saw the BMW VI in-line engines replaced by BMW 132 Dc radials. Only two examples of each version were built. The He 111 V-3 was a prototype for the A-O series, ten of which were sold to China. He 111 V-5 (D-APYS) was the prototype for the B-series which was built for the Luftwaffe from 1936. The pre-production B-O series was used for unit trials, and the B-1 was employed in action in Spain by *Bombergruppe* K 88. The B-2 received more powerful engines and was in large-scale production by the end of 1937. Several of the B-2 version were also employed in Spain. B-O, D-AXOH, served as a test bed for the Junkers Jumo 210 and 211 engines. He 111, G D-ASAR, was fitted with Daimler-Benz DB 600 engines as the V-16. It was intended to be the

The first He 111 B-1 bombers from the factory wore civil registrations. They did not receive military markings until they were delivered to the units.

He 111 B

Left: Two of the first He 111 B-1 bombers on a training flight.

Left below: He 111 B-1 bombers during a flight over Mecklenburg.

Below: Another photo of an He 111 B-1. The photograph was retouched prior to publication, with all weapons being removed.

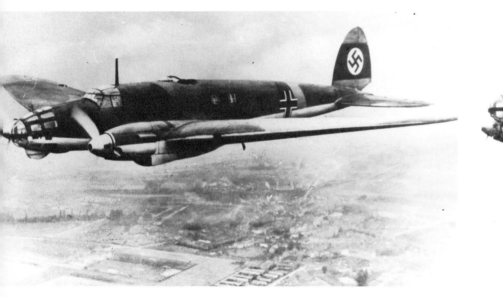

One of the first photographs of the He 111 released for publication outside Germany, this illustration appeared in the English magazine "Aeroplane" in 1937.

A photograph taken during flight testing of the He 111 B: The test team is seen with the sandbags used in load tests.

An He 111 B-1 of the *Legion Condor*'s *Kampfgruppe* 88 over Spain in 1937. The retractable ventral gun position (*C-Stand*) is plainly visible. In later versions this was replaced by a streamlined ventral bath.

Legion Condor

Right: An He 111 E-1 of K 88 during the Battle of Ebro.

Below: An He 111 B-1 of I./K 88. Peter, the *Staffel* mascot, was killed in an air battle over Sagunto.

prototype for the G-4, however this series was not built. The aircraft later became the personal transport of State Secretary Erhard Milch. V-9 (D-AQOX) was the prototype of the D-Series. Only one aircraft, D-AOHA, was built, as this series was abandoned in favor of the E-series. The prototype for the E-Series was He 111 V-10 (D-ALEO). It was similar to the D, but had Jumo 211 engines in place of the DB 600s. The E-1 also saw action in Spain. Versions E-2 to E-5 were built, with the E-3 entering large-scale production. The Heinkel He 111 V-11 was the prototype for the F-Series. It was largely similar to the E, but had a new wing with a straighter leading edge. Thirty F-1s were built for delivery to Turkey, as well as forty F-4s for the Luftwaffe. He 111 V-18 was the prototype for the J-Series. This aircraft, D-A+DUM, was the first version of the He 111 to be fitted with racks for the carriage of torpedoes instead of bombs. It was converted from a B-2 and therefore had the elliptical wing of the earlier version. The J-1, ninety of

Above: He 111 E-1 of K 88 at its home field.

Below: He 111 E-1 of K 88 during a mission against Bilbao's "Iron Ring."

Above: An He 111 E-1 in Germany: its crew is preparing to board the aircraft.

Below: The crew of an He 111 E-1 of K 88 during preflight preparations.

Accidents in Service with Flying Schools

Above: On 1 April 1941 a Ju 52 rammed an He 111 B-2 at Prague-Rusin.

Above: This He 111 B-2 crash-landed not far from Suchdol, near Prague, on 8 May 1940.

Left: This He 111 crashed near Jenec, Moravia on 16 July 1941.

This flying school He 111 B-1 made a forced landing at Prague-Rusin on 3 January 1942.

He 111 V 8, D-AQUO, was a converted B-1 used to test the new asymmetrical nose.

He 111 H-3: In the nose is the observer/bombardier with the forward MG 15. Located in the bulge beneath the fuselage nose is the bombsight. The bomb bay doors are clearly visible on the underside of the fuselage.

An He 111 P-1 during an acceptance flight, escorted by an He 100 D-0 fighter.

He 111 P

Right: This in-flight photograph of an He 111 P-1 shows the appearance of this version as it went into action in 1939. The P remained in front line service only until 1940, when it was relegated to the training schools. Several He 111 Ps returned to the front near Stalingrad in 1942/43.

Left: The characteristic asymmetric shape of the forward fuselage is obvious in this forward view of an He 111 P-1.

Above: A view of the nose of an He 111 P. The bombardier demonstrates the use of the forward MG 15. Visible below him is the bombsight housing.

Following final assembly an He 111 P is rolled out of the hangar at the Heinkel works in Rostock-Marienehe and prepared for its first flight.

which were delivered to the Luftwaffe, was built with the new wing introduced in the F-Series.

All of these *Versuchs* and series machines had a fuselage which was quite similar to the *Lufthansa* version with the exception of the glazed fuselage nose and weapons installations. In 1938 Heinkel's head designer Schwärzler presented a completely new version of the He 111 to the *Technische Amt.* It was the He 111 V-8 (D-AQUO). The machine, an old B-O, had received a completely redesigned fuselage nose. The portion of the fuselage from the pilot's seat to the nose gun position (*A-Stand*) was completely glazed, with curved panels in the upper part and flat panels in the lower. In order to provide the pilot with an unobscured view forward, the *A-Stand* and bomb sight were moved to the right, producing the asymmetrical nose shape characteristic of wartime He 111s. The V-8 served only to test the new cockpit. The prototype for all further production series was the V-19 (D-AUKY). He 111 V-23, a B-2 with the registration D-AHAY, was the test bed for the new ventral gunner's position (*C-Stand*). A ventral bath replaced the retractable "dustbin" gunner's station used in previous versions, which was similar to that of the interim Ju 52 bomber. Although the course of the war saw the He 111 fitted with more powerful engines, increased armament and improved equipment, its basic shape remained the same as that of the V-19. The He 111 V-32 saw the test installation of complete Ju 88 power plants with annular radiators. However, tests revealed no significant advantages over the standard installation.

The new He 111 as represented by the V-19 was to be built in two versions which differed only in the type of power plants installed. The P version was powered by the Daimler-Benz DB 601 and the H version by the Jumo 211. As production of the DB 601 was proceeding at a high rate at the beginning of 1939, the P version was the first to enter series production. The first P-O pre-production machines were delivered to the Luftwaffe for unit trials in the autumn of 1938. The results of these trials were so good that the Supply Office GL/E increased the size of the production order. In the meantime, production of the Jumo 211-powered He 111 H-O and H-1 began at the new Heinkel works in Oranienburg, north of Berlin. The P-series was built primarily at the main Heinkel factory in Rostock-Marienehe and at the Norddeutschen Dornierwerke in Wismar. Production of the He 111 H at Oranienburg soon outstripped that of the P-series, however. Arado Brandenburg as well as ATD Leipzig and the Junkers company also became involved in production. Equipping of the bomber units of the *Luftwaffe* with the new He 111 P and H began in early 1939.

Above: An He 111 P during a test flight over typical Mecklenburg countryside.

Below left: The *Kommodore* of KG 26 in his He 111 H-1.

Below: An He 111 bombardier's view of the aircraft in front.

Employment during the War

On strength with the Luftwaffe *Kampfgeschwader* at the outbreak of war were:

    He 111 E38 aircraft, 32 serviceable
    He 111 J21 aircraft, 20 serviceable
    He 111 P349 aircraft, 295 serviceable
    He 111 H400 aircraft, 358 serviceable

In total 808 aircraft, of which 705 were serviceable. The torpedo aircraft of the J-series were first issued to *Marineflieger* units but were later transferred to the *Kampfgeschwader*. They were soon replaced by H-6 torpedo bombers and relegated to the testing of new bombs, torpedoes and remotely-controlled weapons.

When the Second World War began on 1 September with the campaign in Poland, almost seventy-five percent of Luftwaffe bomber units were equipped with the He 111. Even though the Polish Air Force was numerically far inferior to the German, attacks by Polish PZL 11 and PZL 24 fighters showed that the He 111's defensive armament of three MG 15 machine-guns was too weak. The units improvised, installing additional guns. Some crews wore steel helmets while flying combat missions because the early He 111s had no armor protection. Furthermore, it proved necessary to increase the size of the national markings as, in many cases, German fighters failed to see the small crosses.

The withdrawal of Luftwaffe units from Poland and their transfer back into Germany began on about 20 September 1939. Of these, KG 26, the *Löwen* ("Lion") *Geschwader*, which temporarily consisted of only two *Gruppen* and was led by *Oberst* Sieburg, was placed under the command of the newly-formed 10. *Flieger-Division* under *Generalleutnant* Geissler, whose task it was to carry out the air war against England in cooperation with the

Close-up view of an He 111 P-1 of KG 26. The distinguishing features of the P-Series are visible on the engine cowlings: the supercharger air intakes on the left side of the engines.

He 111 P, V4+ AU, of *Kampfgeschwader* 1 as it releases its bombs over France.

The Bombs are Loaded

Above: Armament technicians carry SC 50 bombs to the aircraft.

Right: Two men carefully push an SC 50 into the vertical bomb cell of an He 111.

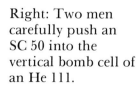

Above: Loading of bombs into the aircraft had to be done with great care.

Kriegsmarine. The *Geschwader* was especially well-suited to this task since most of its members had come from the *Kriegsmarine,* including *Oberst* Sieburg. Geissler's assignment was a difficult one, because how much could he be expected to achieve with sixty He 111s? Nevertheless, on 26 September nine He 111s of KG 26, together with four Ju 88 A-1 bombers of I/KG 30, flew the first attack against British fleet units in the North Sea. For the time being KG 26 remained the main actor on the stage of the air war against England.

In 1940 the He 111 H-3, deliveries of which had begun in November 1939, was used in action for the first time during the Norwegian Campaign. During this time KG 26 specialized in attacks against shipping, achieving considerable success. In the same year deliveries of the P-4 series began. The H-3 and the P-4 were distinguished mainly by an increased armament of four to six MG 15 machine-guns, a 200 kg increase in bomb load and armor protection. In addition to KG 26, KG 4 and *Kampfgruppe* 100 were also equipped with the He 111 for use in Norway. With the P-6 series production of the P variant came to an end. From then on only versions of the H-series were built. Up to that time only the P-4 and H-3 had entered large-scale production; the other versions being built only in limited numbers. The H-5, which differed from the following H-6 series only in the use of VDM propellers in place of the usual Junkers controllable-pitch propellers, was likewise built in small numbers. The H-6, on the other hand, was to become the most produced version so far. It was employed as a torpedo-bomber as well as a regular bomber. I/KG 26, which was stationed at Banak and Bardufoss from 1942, achieved outstanding success against the Allied PQ-QP convoys delivering supplies to the USSR, although losses were heavy. Also worthy of mention are the activities of KG 4 *"General*

Above: He 111 P bombers of *Kampfgeschwader* 55 during the Battle of Britain. These machines have had their undersides and fuselage sides painted black for night bombing missions.

Below: The crew of this He 111 appears to have been less than satisfied with its accommodations. (*Sauladen* = pigpen).

He 111 P as Long-Range Reconnaissance Aircraft

Above and below: He 111 H-2, T5 = BU, of the *Luftwaffe* Commander-in-Chief's Long-range Reconnaissance *Gruppe* (Aufkl.Gr./Ob.d.L.), which was also known as *"Gruppe Rowehl."*

Wever" following the conclusion of the French Campaign of 1940. This *Geschwader* and the units flying He 115 floatplanes laid mines off the southeast and east coasts of England, especially in front of river estuaries, resulting in considerable losses to British sea traffic. The following He 111-equipped units were saw action during the intensified air war against England, the so-called "Battle of Britain": *Luftflotte* 5: KG 26, *Luftflotte* 3: KG 55, KG 27, *Luftflotte* 2: KG 1, KG 53, KG 4 and *Kampfgruppe* 100. The latter unit served as a so-called "pathfinder" unit. Its task was to mark targets with smoke-bombs or parachute flares.

The bulk of the He 111 units committed to the attack against the Soviet Union in 1941 were equipped with the H-6 variant. The units involved were KG 27, KG 53 and KG 55. By this time a large proportion of the bomber units had already converted to the Ju 88. It was at this time that the He 111 became a "maid of all work." A modification of the H-3, the H-8 had first appeared during the Battle of Britain. Thirty H-3 and H-5 aircraft were equipped with a large balloon-cable-cutter array. The resulting machines were very nose-heavy and did not prove a success. The structure was removed from the surviving aircraft, which were fitted with *Rüstsatz* R 2, equipment for towing cargo gliders. A further development was the night bomber H-10, which was built in small numbers in 1942. It was employed over England and had a so-called *"Kuto-Nase"*, an electrical cable cutter, installed in the leading edge of the wing. An improved version of the H-10 was the H-11, which had strengthened armament and armor and could be employed as a bomber or glider tug.

In autumn 1941 KG 4 was transferred to the northern sector of the Eastern Front and placed under the command of *Luftflotte* 1. It was there that the unit's H-6 aircraft were employed for the first time not as bombers, but as transport aircraft. When, on 8 February 1942, the German II. *Armee-*

Above: Front view of an He 111 H-1, powered by Jumo 211 engines.
Below: Dorsal gunner's position (*B-Stand*) with hood open and MG 15 deployed.

Above: Dorsal gunner's position with hood in the closed position.

Ventral bath of an He 111 H-1 with entrance hatch and *C-Stand* (MG 15).

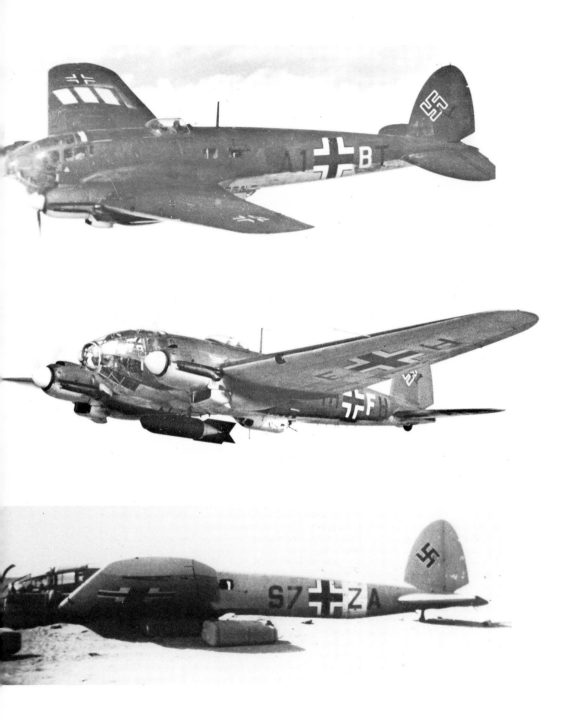

Above: Bombardier at work: A good view of the *"Lotfe"* bombsight. To the left from top: He 111 H-3, A1 = BT, of KG 53, He 111 H-6, 1H + FH, of KG 26, belly landing by an He 111 H-5trop used as a transport by the *Stab* of *Stuka-Geschwader* 3.

One of the first He 111 H-5 bombers converted to carry torpedoes.

After its practice torpedoes have been loaded the He 111 H-5 is towed to the runway. The later He 111 H-6 was built from the outset to carry bombs or torpedoes.

Above: The pilot of an He 111 H-6 briefs his crew before a mission.

Below: In the foreground an SC 1000 high-explosive bomb; in the background is an SC 500.

Above: SC 50 bombs are removed from their shipping crates.

Below: An He 111 *Staffel* carries out a daylight attack during the Battle of Britain.

He 111 H-6

Above: He 111 H-6 of KG 26, 1H + DN. The white fuselage band was worn by Luftwaffe aircraft operating in the Mediterranean Theatre.

An He 111 H-6 shortly after takeoff from Oranienburg for the front.

He 111 H-6, A1 + KL, of KG 53. A large proportion of the H-6 series carried a fixed, rearwards-firing MG 17 in the aircraft's tail cone.

korps and elements of the **X.** *Armeekorps* were surrounded in Demyansk by the Soviets, the He 111s of **KG 4** were the first aircraft to fly into the pocket, dropping supply canisters to the hard-pressed 123rd Infantry Division. Following heavy losses to the Ju 52s of **KGrzbV 172**, the He 111s of **KG 4** also had to take over the job of flying supplies into the smaller, neighboring Cholm Pocket.

A new version of the He 111, the H-16, entered large-scale production in the autumn of 1942. It was to be the last major production series of the He 111. The H-16 was conceived for operations on the Eastern Front where Russian defences were weaker than in the West and where German bombers were often forced to fly without fighter cover due to limited resources. It had a five-man crew and an extremely powerful defensive armament: *A-Stand* (nose): 1 **MG/FF** (occasionally replaced by an MG 151/20), *B-Stand* (dorsal position): 1 **MG 131**, *C-Stand* (ventral position): MG 81 Z or MG 131, 2-4 MG 15 or MG 81 firing from fuselage windows. Maximum bomb load (short range) 3,250 kg, rangewith 1,000 kg of bombs 2,900 km. Using various modification kits (*Rüstsätze*) the H-16 could be employed as a pathfinder, night harassment bomber or glider tug. Several He 111 H-18 bombers were used by **KG 40**, which operated against shipping targets in the Atlantic, equipped with FuG 200 search radar.

The He 111 H-20 entered service in 1944. Use of modification kits permitted the aircraft to be employed as a transport for 16 paratroops, a glider tug or as a night or harassment bomber. Several H-21s were built in 1945. Powered by the 1,750 h.p. Jumo 213 E-1, the H-21 could reach a maximum speed of 480 kph. The He 111 B-2 of 1937 had attained 370 kph!

The He 111 H-12 and H-22 were special-purpose versions. The former was employed to drop remotely-controlled Hs 293 and Fritz-X

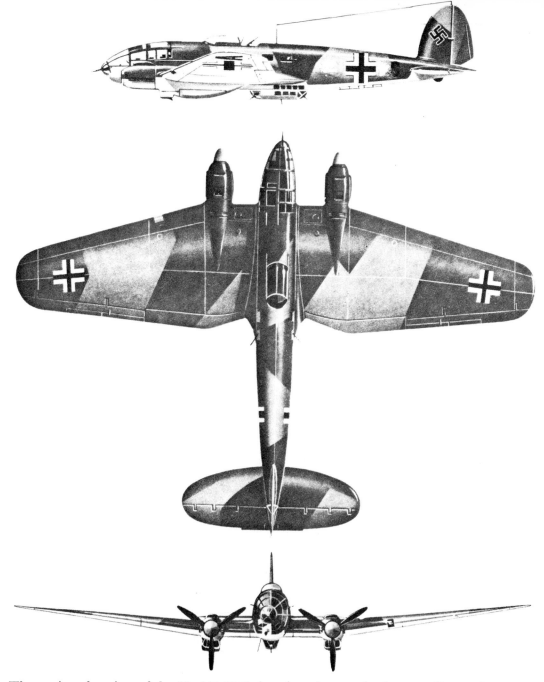

Three-view drawing of the He 111 H-6 showing the standard camouflage scheme. The colors were Dark Green 70 and Black Green 71 above, and Horizon Blue 65 below.

He 111 H-11 with nose-mounted MG/FF cannon.

He 111 H-5 with external bomb load, an SC 500.

An SC 1000 high-explosive bomb is loaded onto an He 111 H-6.

One headache for Heinkel designers was the He 111's wing/fuselage filet.

Above: An He 111 H-8 with balloon-cable fender and cutting gear, which was later removed.

On 12 February 1941 *Generalleutnant* Rommel, who had been named commander of German forces in Africa, arrived in Tripoli aboard an He 111 H.

He 111 H-2, 5K + GA, of KG 3 with Fl 103 missile (also FZG 76 or V 1).

He 111 H-12, OF + OV, as trials aircraft for testing the Bv 143 B missile (a steerable, rocket-powered surface torpedo).

Above: He 111 H-6 of KG 26 in the Mediterranean Theatre.

Above right and right: In overloaded condition (3,000 kg of bombs) the He 111 required the use of so-called "R-Geräte" (takeoff assistance rockets) to become airborne. Here an He 111 H-6 has just ignited the "R-Geräte" and is beginning its takeoff roll.

Supply canisters such as these (above), the so-called "supply bombs," were dropped by He 111s to Army units in the East which had been surrounded or cut off.

Below: An He 111 crew which has just completed its 500th supply mission.

Above: External bomb racks had to be used to carry heavy loads (large-caliber bombs or supply canisters).

Three Knight's Cross wearers who flew the He 111.

An He 111 H-16 of KG 26. This *Geschwader* saw action over England, the North Sea and the Mediterranean.

The He 111 H-5 was the first version with two PVC racks beneath the fuselage, as seen on this aircraft.

Oberfeldwebel Baumann (right) was among the first He 111 pilots to be awarded the Knight's Cross.

He 111 H-16 during winter operations in Russia. The aircraft could carry as many as five SC 250 high-explosive bombs beneath the fuselage.

Above: He 111 H-16s of *Kampfgruppe* zbV 100 during the aerial supply of Stalingrad. Below: An H-6 as seen from the aircraft behind and above.

bombs, while the H-22 was a launcher for the Fi 103 flying bomb, also designated "V 1", against targets in England. The final version of the He 111 was the H-23. Originally intended as transports for eight paratroops, the aircraft were converted into bombers.

The high-altitude bomber R-Series failed to reach production. The previously mentioned V-32 was to have served as the prototype for the series. After initial tests with a normal Ju 88 engine were successful, the aircraft was fitted with the DB 601 U with TK 9 supercharger and annular radiator. Development was then dropped.

Finally, one special development must be mentioned which, although an improvisation, proved to be the best solution for the desired purpose. Following construction of the large-capacity Messerschmitt Me 321 V-1 glider, the question of a suitable tug aircraft became a problem. There were not sufficient numbers of the Ju 90 available, and this aircraft was the only one capable of towing the Me 321. The first solution, the so-called *"Troika-Schlepp"* with three Bf 110s, proved to be life-threatening during initial tests in Orsha on the northern sector of the Eastern Front. *Generalluftzeugmeister* (Director General of Equipment) *Generaloberst* Ernst Udet therefore rejected the *"Troika-Schlepp."* While visiting Heinkel he inquired whether it would be possible to create a single aircraft from two He 111s — that is a four-engine tug aircraft —while keeping costs low. Siegfried Günter, head of the project office, and Schwärzler, the chief designer, got together afterwards and proposed a five-engine He 111. This consisted of two He 111 H-6 bombers whose left and right wings respectively were removed and joined by a new wing centre section which housed a fifth engine. The first two He 111 Z, as the new aircraft was named, were ready in late autumn 1941 and no difficulties were encountered during tests. Heinkel received a contract for ten He 111 Z glider tugs.

Above: Returning from a night bombing mission in the East with battle damage, this He 111 H-6 of KG 53 (A1 + KL) made a forced landing.

Right: SC 500 and SC 1000 HE bombs wait to be loaded aboard an He 111 H-4.

He 111 Z

The He 111 Z/Me 321 combination was to have been used in the conquest of Malta. The planned operation did not, however, take place. The He 111 Z first saw action in 1943 during the Stalingrad airlift. The first two Me 321 glider trains landed at the planned forward base in Majewka on 28 January 1943. These saw no further action, however, as the Stalingrad airfields of Pitomnik and Gumrak had meanwhile been captured by the Soviets. Later a total of eleven glider trains were employed in the supply of the Kuban bridgehead where they proved very successful. Later the Me 321 was replaced by two Go 242 cargo gliders as these had proved more suitable. The glider trains were also used in the battle for Italy. Eight of the He 111 Z glider tugs were lost in air attacks. By autumn 1944 four were left.

Proposed developments of the He 111 Z were the He 111 Z-2 long-range bomber and the He 111 Z-3 long-range reconnaissance aircraft. These designs progressed no farther than the drawing board, however.

A total of 5,656 He 111s was built from 1939 until the time production ceased.

Forward view of an He 111 Z-1. Four of the aircraft's five propellers are visible as well as the nose of one of the two He 111 H fuselages which made up the aircraft.

Above: He 111 Z-1 in flight. Below: For long-range missions the He 111 Z-1 employed four 300 liter external fuel tanks.

The bitter end: At war's end hundreds of brave He 111s lay scattered about Europe. Some had been shot down, some destroyed on the ground or blown up by the Germans themselves.

This He 111 was blown up by its crew at Prague-Rusin airfield.

This He 111 H-6 survived the war and was taken to the USA, where it was flown several times.

Technical Data for the Major Variants

Aircraft Type	He 111 B-2	He 111 C	He 111 P-2	He 111 P-4	He 111 H-6	He 111 H-16	He 111 H-21	He 111 Z
Role	Bomber	Verkehrsflz.	Bomber	Bomber	Bomber	Bomber	Bomber	LS-Schlepper
Crew	4	2 + 10	4	5	5	5	5	7
Power Plants	DB 600 CG	BMW VI 6, oZU	DB 601 Aa	DB 601 A-1	Jumo 211 F-1	Jumo 211 F-2	Jumo 213 A-1	Jumo 211 F-2
Horsepower	2 x 950	2 x 660	2 x 1020	2 x 1100	2 x 1175	2 x 1350	2 x 1775	5 x 1350
Wingspan (m)	22,60	22,60	22,60	22,60	22,60	22,60	22,60	35,20
Length (m)	17,50	17,50	16,40	16,40	16,40	16,40	16,40	16,38
Height (m)	4,40	4,10	4,02	4,00	4,02	4,02	4,02	4,53
Wing Area (sq.m)	87,60	87,60	86,50	87,60	86,50	86,50	86,50	147,0
Weight Empty	5840	6210	8015	8015	8020	8680	10500	21400
Gross Weight (kg)	8600/10000	8390/9600	13300	13500	13300	14000	16000	28400
Pay Load	-	3390/2180	-	-	-	-	-	6930
Maximum Speed (kph)	370	315	400	398	400	410	480	435
Crusing Speed (kph)	345	305	360	373	370	370	405	392
Landing Speed (kph)	120	110	115	115	115	115	125	120
Ceiling (m)	7000	4800	7400	8000	8000	8500	10000	10000
Range (km)	910	1000/2200	960	1970	2500	2100	2900	-
Take-off Distance	520	510	1050	1120	1200	1150	1200	-
Landing Distance	600	600	950	1000	1000	1000	1000	-
Armament	3xMG 15	-	3xMG 15	6xMG 15 1xMG 17	1xMG/FF 5xMG 15 1xMG 17	1xMG/FF 1xMG 151 4-6 MG 81	3xMG 131 2xMG 81 Z	2xMG 131 8xMG 81
Maximum Bomb Load (kg)	1500	-	2000	2800	2500	3250	3000	-

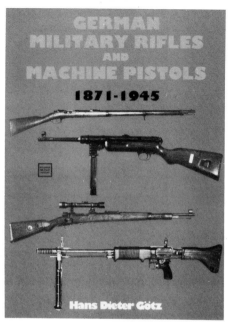

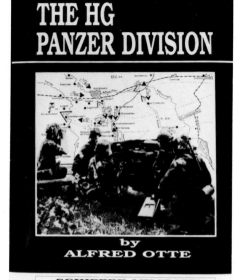

• *Schiffer Military History* •

Specializing in the German Military of World War II

Also Available:

• The 1st SS Panzer Division - *Leibstandarte* • The 12th SS Panzer Division - *HJ* • The Panzerkorps *Grossdeutschland* •
• The Heavy Flak Guns 1933-1945 • German Motorcycles in World War II • Hetzer • V2 • Me 163 "Komet" •
• Me 262 • German Aircraft Carrier *Graf Zeppelin* • The Waffen-SS - A Pictorial History • Maus • Arado Ar 234 •
• The Tiger Family • The Panther Family • German Airships • Do 335 •
• German Uniforms of the 20th Century - Vol.1 The Panzer Uniforms, Vol. 2 The Uniforms of the Infantry •

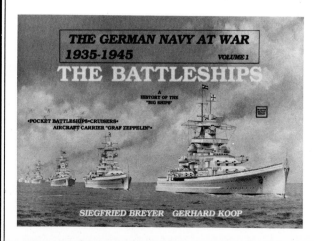